# The Jazzalot Journal

Between the Books – Volume 1

# Business Infrastructure Overview

## Jump-Start / Maintain Control of Your Indie Publishing Business

## LaRae Schmitt

Jazzalot Books

# Many Thanks

**My Sincere Gratitude** to ALL those who work to provide endless information on the Web. The tools, opinions and instructions that have helped me in my journey — have been nothing short of AMAZING.

**To ALL OPEN SOURCE AND FREE SOFTWARE / TOOL Developers / Experts / Organizations** — You All Are Some of The Greatest Unsung Heroes of Our Time!

Sharing my experience both in this format and on my Blog Series "Between the Books" is one way I plan to

## 'Pay It Forward'

!
Making
the
End
Product
Possible
!

# Table of Contents

# Chapter 1
# Business Support

## A Little Bit of Structure – Goes a Long Way

This book is meant to be used as an outline for points of interest you might wish to implement – and a workbook to assist in creating / streamlining your own business.

- In each chapter – there is an 'overview' followed by examples of

the actions I have taken to create my own book-publishing business.

- Each chapter provides an additional area for you to write your
own notes as you proceed.

## Notes – Notes – Notes

It is important to keep notes from the beginning – for highlights and
reminding yourself of information, ideas and things you have already
resolved – prior to adding even more complexity to the issue (which
incidentally MAY change your resolve) – and ultimately freeing your
mind to move forward.  Simple but Effective...

- Keep a paper notebook – for a 'scratch pad'.

- Purchase notebooks during school-supply season to stock up at bargain prices – giving 'permission' to use freely.

- Put the book / project name on the front and that becomes your go-to for 'lists', brainstorming, etc., for that subject in general.

## Digital Notes

Log your final information in your own digital notes for ease of finding – and so that you will have to figure it out only once.

- Later, when working on 'metadata' for website, Search Engine Optimization (SEO), book metadata / identification... having this information in digital format will allow you to simply search/cut/paste as needed.

- This helps to keep all description data the same – which is said to increase your chances of being 'found' by the internet and customers.

- Taking time to document Tips and Tricks learned can also be handy for quick reference, months down the road.

## Folder/File Naming, etc.

- Folder/File Naming, Organization, Backups, etc., are Paramount.

- You can save often (at least once/day or once/hour) by simply using the 'save-as' and adding a digit to the end of your filename. This way you can always go back.

- It only takes redoing something 100 times to figure this out –

and you will thank yourself often for implementing a 'system'. Anything is better than nothing.

## Business / Domain / Social Media Name(s)

It is important to decide what you will call your publishing company, etc., as this can help to build your 'business perception' and will be used from now on. If you choose to have a website or depend solely on social media will be a personal decision.  Here are a few things to consider:

- Purchasing domain name(s) for your company can start to build ownership history and serve to strengthen your 'brand' on the internet.

- Deciding a business name can be affected by what domain and/or social-media name(s) are available.

- Great Domain Names are usually catchy, short, easy to spell, recite and remember.

- Checking the history of domain names, similar words or unique spellings — can help you decide which one(s) to choose.

- If you do decide to purchase domain names, I would purchase them as soon as your research is complete — so that they are still available. Also establishing brand-specific email / social media accounts, e.g.

- Purchase a domain name for your Publishing Company.

- Purchase a domain name for your Imprint name.

- Purchase a domain name for your Pen Name.

- Create your business email and social media accounts.

## Set up Business License, etc.

- Set up your City, State and Federal business licenses / numbers per your area requirements.

- Note: When you create your account for print / distribution — at some point they will require you to enter 10-99 information for profits, etc.

# Acquiring Domain / Web Assets

All my domain/web services are purchased from godaddy.com, as I have done so off-and-on for years. They are a solid company with 24/7/365 Customer Service – and more than once have walked me through new/complex areas to change or implement what I needed.

- They offer a full tool suite to choose from – making it pretty much a one-stop-shop for your online presence, direct sales infrastructure, blog and social media tools, etc.

- They provide 2-factor security when logging in to make changes, which I always take advantage of.

- They also have impressive AI products and support.

- All of this, along with their Video Training, a-la-carte and progressive tool offerings, provides assurance that my business can maintain access to a robust, integrated, online support system.

# My Business / Domain-name Setup, etc.

- **Jazzalot.com** – reflects the main business name / publishing company name. This is used as the main website, for selling my books and where other products can be sold as well. The name is general enough to support selling products as a 'general store'.

- **JazzalotBooks.com** – reflects the 'imprint' name that is used on my books. This domain name is redirected to jazzalot.com, so when anyone types the imprint name, they will be redirected to the main website – that will show all books /  products and links to their particular item of interest, etc.

- **LaRaeSchmitt.com** – keeping this domain name for now as it reflects my 'pen' name. This domain name is also redirected to jazzalot.com, so that when anyone does a web search – it will show up as jazzalot.com (general store).

- Redirecting domains to the main domain should also provide cross-referencing information for web-crawlers / SEO...

- It is suggested to always refer to your domain names in lower case – for web purposes. I try to do this as much as looks 'OK' – but often defer to initial caps for added clarity.

- **JazzalotBooks@<email service provider>** I chose to have a book-related email, as it can further solidify your book business under your imprint name. Some companies are said to use this info to 'catalog' your books and I wanted any such cataloging to point to my publishing company imprint name – Jazzalot Books.

- **JazzalotCS@<email service provider>** future Customer Service.

# Chapter 2

# ISBNs / Barcodes / QR Codes

## How – What – When – Where – Why

Each ISBN number is unique per product type. For one book, it will take a separate ISBN number for the hardback copy; another ISBN for the paperback; and yet another for the ebook version.

This is the number that tracks locating and ordering of the book –

pretty much for the life of that version.

## Free ISBNs

You can often get free ISBNs from the print-on-demand (POD) company (my POD company is IngramSpark.com).

- Just know, that the 'imprint name' will be theirs and not yours.

- Ingram Spark's 'imprint' for free ISBNs – shows up as 'Indy Pub'.

- 'The world', through distribution, will see your book as having the POD company 'imprint' (check for current details online).

- You will not have control to move that 'version' away from that POD company.  It will always be partially owned by them, etc.

- If you wish to move it – that would require uploading as a new title with your own ISBN (always check online for specifics).

## Purchased ISBNs

This is the one place I would spend money – at least for your 'corner stone' products.

- Purchased ISBNs reflect your own publishing company / imprint name – which helps to promote your 'brand'.

- You have full ownership of the product.

- You can move the book to any printer if/when...

- Since you own your ISBN – and as of my signing of the Ingram Spark contracts – they allow you to move your book to another company with short notice and no strings attached (This is where it is important to beware of the contracts you sign – as  not all POD companies have been known to allow this as readily).

- Bowker (Bowker.com redirects to myidentifiers.com for ISBN purchases). Bowker is the OFFICIAL ISBN Agency for publishers physically located in the US and its territories...

- Working with Bowker is very easy and they have a phone number for quickly resolving any tough questions.

- You can purchase ISBNs in bulk, reducing the overall cost per ISBN – and they do not expire.

- ISBN's stay organized in one account that is easy to use and update your information.

- Bowker, as part of this process, and at no additional cost, adds your book information to their *Books In Print*® database used by thousands of Retailers, Libraries, Schools, etc.

- BEWARE of purchasing ISBNs from anyone but an official ISBN Agency.

# Barcodes

Barcodes can be purchased as well, however, I would stick to the one provided by the Print-on-demand company (if applicable).

- Barcodes are provided for free from the print-on-demand process with Ingram Spark.

- With Ingram Spark, a barcode comes as part of your cover template with instructions on how it can be moved.

- It automatically meets technical specifications.

- NO NEED TO PURCHASE!

# QR Code

It seems that QR Codes can be readily made for just about anything, however, this one I prefer to Purchase.

- QR Code – purchasing this from Bowker.com for a nominal price, keeps it with a reputable company.

- Bowker keeps it within your account, along with your list of purchased ISBNs.

- The QR Code they provide – allows the pointer/location to be changed, to point to a different domain....

- So that years later, if you change your domain name – the QR codes printed on 'existing' books – will redirect to your 'updated' domain name automatically.

## How I Currently Use ISBN / Bar & QR Codes

I Purchase ISBNs for my books from Bowker.com (aka myidentifiers.com) using my jazzalotbooks email address for my username.

- Purchased ISBNs in bulk for various titles/formats... (bulk can reduce the price-per-ISBN to ~ $25 and less – check their website).

- Use Bowker ISBNs for all Children's Books.

- The original plan was to use Free ISBNs for *The Jazzalot Journal* since the journaling will be more informal information sharing, with periodic version updates... However, due to the added 'branding' of using my own company / imprint name – I decided to use my own ISBNs, purchased direct from Bowker.com.

- A Barcode for each book comes with the cover template (free from ingramspark.com) and I move it within the given parameters, to fit nicely with my cover art.

- QR Code – purchased only ONE QR Code from Bowker.

- Have the QR-code definition (URL) at Bowker, pointing to the main business website with https://www.jazzalot.com as the address (Note: be SURE to include the 's' on the end of the http).

- It is important to make sure the QR Code Address, at Bowker, includes https:// before the domain name – so it will show up as a 'secure' location – when read by electronic devices, phones, etc...

- Placing QR code on Back cover and inside copyright page of all books (actually, after my first – as this is where I learned to also use the copyright pg). Note: some/all 'ebooks' do not include the back cover of a book – so also placing the QR code on the copyright page, increases the chance of it being seen and accessible.

- Displaying this same QR code on ALL products – brings them to the main website / 'general store' landing page.

- As you grow – different pages can be created for different products – all within that same website.

- Less overall cost / complexity.

# Chapter 3
# Software / Tools

## Overview / Software Tools Downloaded

My focus was to stick with 'free and open source' software to minimize overall cost, etc. – and to only select tools for local install (not on the cloud). My research consisted of reading / listening to lots of opinion, keeping notes, etc. - until one tool gets mentions from multiple sources and then drilling down from there. These are the software tools that I downloaded for my book creation and they are all incredible!

## Notes / Writing – LibreOffice (libreoffice.org)

This 'Free and Private' office suite (successor to Open Office) is used by millions. It includes Writer(word processing), Impress(presentations), Draw(vector graphics and flowcharts), Calc(spreadsheets), Math(formula editing), and Base(databases).

- The document Writer is great for writing original rough-draft text for books – prior to copy / export to Scribus (final 'book layout').
- Use document Writer to keep electronic notes – one file for each major area of notes comes in extremely handy, etc.
- Compatible with other similar programs and can export in many/all of the most popular formats.
- Offers export to PDF Format – perfect for sharing files in final format (great for 'uniform' display on most/all devices).
- The Impress presentation program works great to compare screen-shots of book illustrations, to visualize competing designs, mock-up layouts, etc.

# Painting / Illustration – Krita (krita.org)

Krita is a powerful illustration tool – that also has animation capabilities... You can illustrate all electronically or import your original art to further modify / manipulate to your satisfaction – for both Book and Cover Art.

- There are fantastic tutorials online where you can gain answers to just about any question you will come up with.  Like usual – sometimes the trick will be determining just HOW to ask the question.

- This tool provides endless capabilities and I continually discover new shortcuts, etc.

- Note:  Exporting to PNG format is recommended – for importing artwork to Scribus / final 'book layout'.

# Desktop Publishing – Scribus (scribus.net)

This tool works great for the actual layout of your book – as it will appear in print. Scribus makes it easy to 'visualize' exact placement of your pages, artwork, etc.

- I also use this in the initial (picture) book design phase to rough out the words per page with additional text blocks for notes describing the illustrations desired, and how they will fit with the words, etc.

- There are loads of online tutorials for pretty much anything you will need with this tool. Once you learn the basics, it will be your 'go-to' – and learning new tricks will also reveal the exacting control it allows.

- Scribus exports to 300dpi PDF/X-1a required for printing – and takes care of other necessary details.

- My files upload to the POD company without errors and print beautifully.

- Scribus is also great for brochures, magazines, etc.

## Fonts – OpenDyslexic (opendyslexic.org)

Fonts are interesting. Like Software, you will likely have to pay for a 'license' to use them – OR find a Free / Open Source Font that you like. I found OpenDyslexic (which is what I am using in this printed journal).

- I wanted my children's books to be larger print (I use 18-point minimum) – easily read by low-vision and dyslexic readers.

- OpenDyslexic Font is ranked as one of the best 'accessibility fonts' for dyslexia / visually-impaired audiences.

- Plus, I would be hard-pressed to find a more suitable-looking font for my Children's books!

- Note: Make sure that the 'Free' Font(s) you choose have licensing that confirms that they are totally – Free/Open Source.

# Chapter 4
# Online Tools / Services

## Online Services

My goal is to work with online services that are solid, with decent customer support. Also, like above, reading many reviews, etc., along with past experience – these are the services I chose – and they work Great!

# Print-on-Demand – Ingram Spark  (ingramspark.com)

Ingram Spark is an indie publishing company set up (as a boutique company of a much larger parent company) to give indie publishers the tools and assistance needed to upload, print and distribute books/ebooks to over 40,000 retailers globally.

- Your book will be accessible by retailers, although being new, you may have to search for the ISBN -vs- the title at times.

- They provide easy print-to-ebook conversion, both through their free tool or conversion service (check online for pricing).

- This makes Ingram Spark pretty much a one-stop print / ebook distribution company.

- Your account includes a portal for managing your print and ebook products, pricing, orders, sales reports, etc.

- NOTE: Whatever print-on-demand company you ultimately decide to go with, I would suggest that BEFORE SIGNING ANY CONTRACTS making distribution possible – IT IS HIGHLY RECOMMENDED THAT YOU CAREFULLY READ / HIGHLIGHT AND COMPARE the contracts of any / all major indie-publisher print-on-demand companies you may be considering and ACQUIRE LEGAL COUNCIL AS YOU DEEM NECESSARY.

# Overview of My Ingram Spark Account Setup

Created account, using jazzalotbooks email as my username, and filled out 10-99 / business information so that they can send earnings to the designated location, which can include PayPal, etc.

- Signed contract(s) with IngramSpark.com for my Print-On-Demand and Distribution.

- Signed the additional contract(s) for all major retailers who provide addendum / contracts to the Ingram Spark portal, etc.

- Although they offer a paid assistance, I use the online contact (usually respond within a few days) and look through their extensive help library for any assistance needed.

- Downloaded their latest "File Creation Guide" in PDF format – that provides answers for setting up and uploading your book for print / distribution.

- Upload book and input book pricing (distribution is free up front – as they take payment out of proceeds).

- Note: IngramSpark.com is an online service for 'formal and final' version of a printed book / tracking sales / etc., so any inputs are treated with gravity. Once finalized, making inadvertent 'changes' to things like the 'metadata' may trigger some type of error... I usually address any 'errors' / give it time to 'settle out' / back out changes / re-upload files or delete and start over.

- The above also underlines the importance of having your 'metadata' in your electronic notes — ready and edited to your liking — so that you need only 'cut and paste' into desired fields on these end-product sites...

## Website / Tools – GoDaddy (godaddy.com)

GoDaddy has many tools for pretty much anything you wish to create as a website, sell, advertise, manage communications with your customers, etc.

- They offer 'AI' and can even set up your website for you. However, their help files, 24/7/365 live support has always been impressive and their tools pretty straightforward so I chose to easily create my own.

- Taking a few minutes to learn what you need, allows you to simply update your website without the wait time that might otherwise be necessary when having someone else do it.

- I use my personal email address as my username for logging into this site, as the domains / services purchased can also be for past and future (multiple) businesses.

## My Current Website Setup with GoDaddy

GoDaddy's Websites + Marketing is my 'go-to'. I currently have a basic website layout and a "free activities" page, where I add a new activity per month. The older activities drop off and are saved for a future "activities" book. *The Jazzalot Journal* is my blog page – data from which can feed into future 'how-to' books, etc…

- I originally displayed links to my book(s) with five-or-so popular retailers. Now I refer to the retailers in general – but also put "Share & Sell" links that Ingram Spark provides – where people can order directly through them at a discount ( and the profits come back to the designated account, similar to any other retailer transaction).

- The QR code on the back and copyright page of each book directs people to jazzalot.com. Under the QR Code on the back of each book, it also reads "Visit www.Jazzalot.com for Activities and More!". This is the only 'advertising' I started out with.

- When researching – they argued to keep writing as a form of advertisement… as spending hours on saturated social medias may or may not return great results, but the time spent writing another book at least gets you another book...

- Relying mostly on the QR Code on the books with message re: "free activities" ('call' to the website), plus the thousands of retailers who have access to the books through Bowker and Ingram Spark – has provided sales that have continued to trickle in since the beginning.

- After the first year or two, I did get my book into the local library (by simply asking them to order it) and told a few stores at the mall, etc., however, my plan is to publish a few more books prior to venturing into the advertising world...

- My new Blog (*The Jazzalot Journal* 'Between the Books" Series at www.jazzalot.com) is a long-awaited step towards 'Paying It Forward' – that turned into being a product – which is this book, along with future volumes planned...

- As indicated above, 'writing another book' can ultimately turn into a form of advertising, with options that can ultimately provide You more control.

- "One Step at a Time" —  and similar old adages, are useful to remember — as you focus on building Your Next Success!

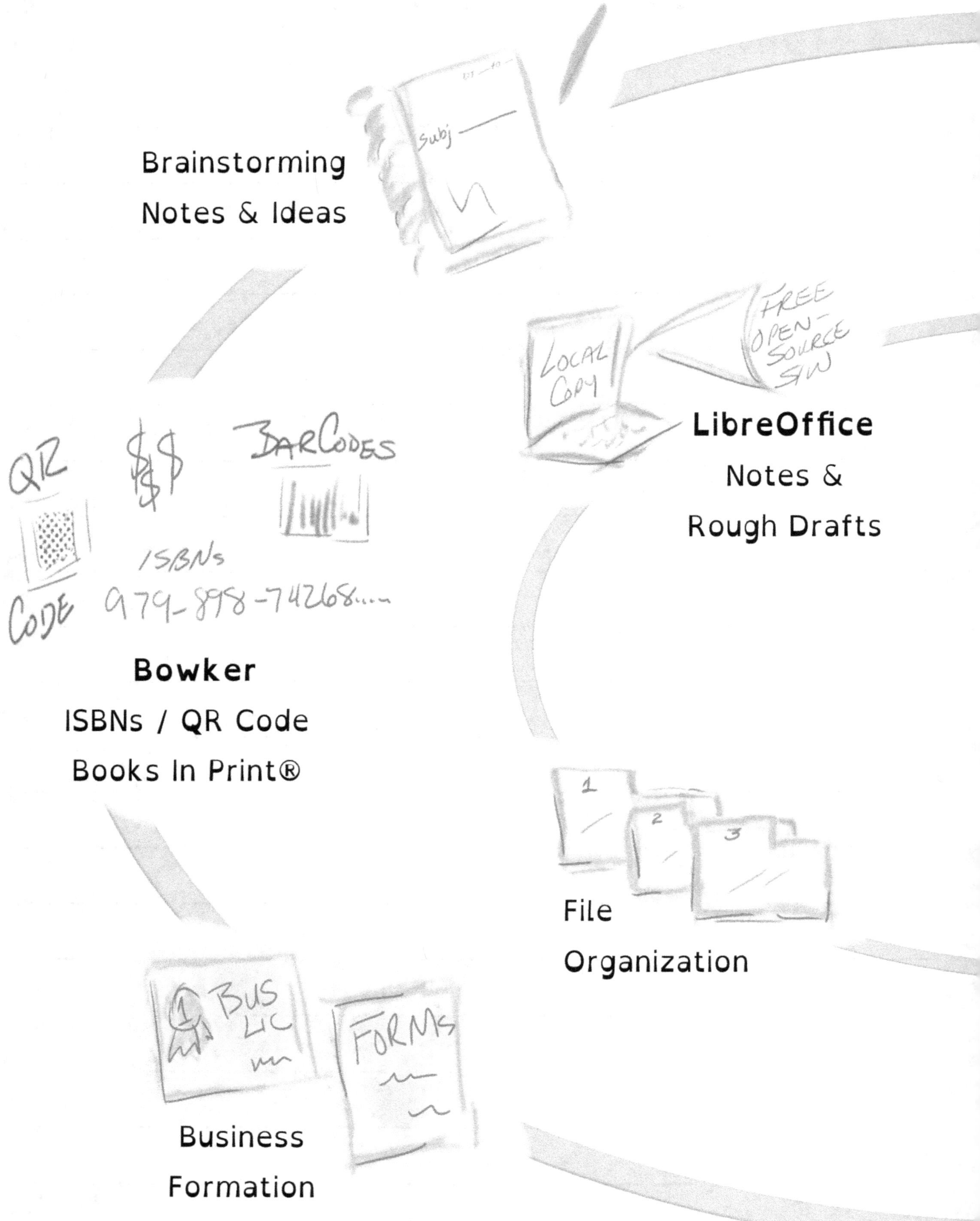

Brainstorming
Notes & Ideas
Subj
LOCAL COPY
FREE OPEN-SOURCE S/W
LibreOffice
Notes &
Rough Drafts
QR
$$
BARCODES
ISBNs
CODE
979-898-74268....
Bowker
ISBNs / QR Code
Books In Print®
1
2
3
File
Organization
BUS LIC
FORMS
Business
Formation

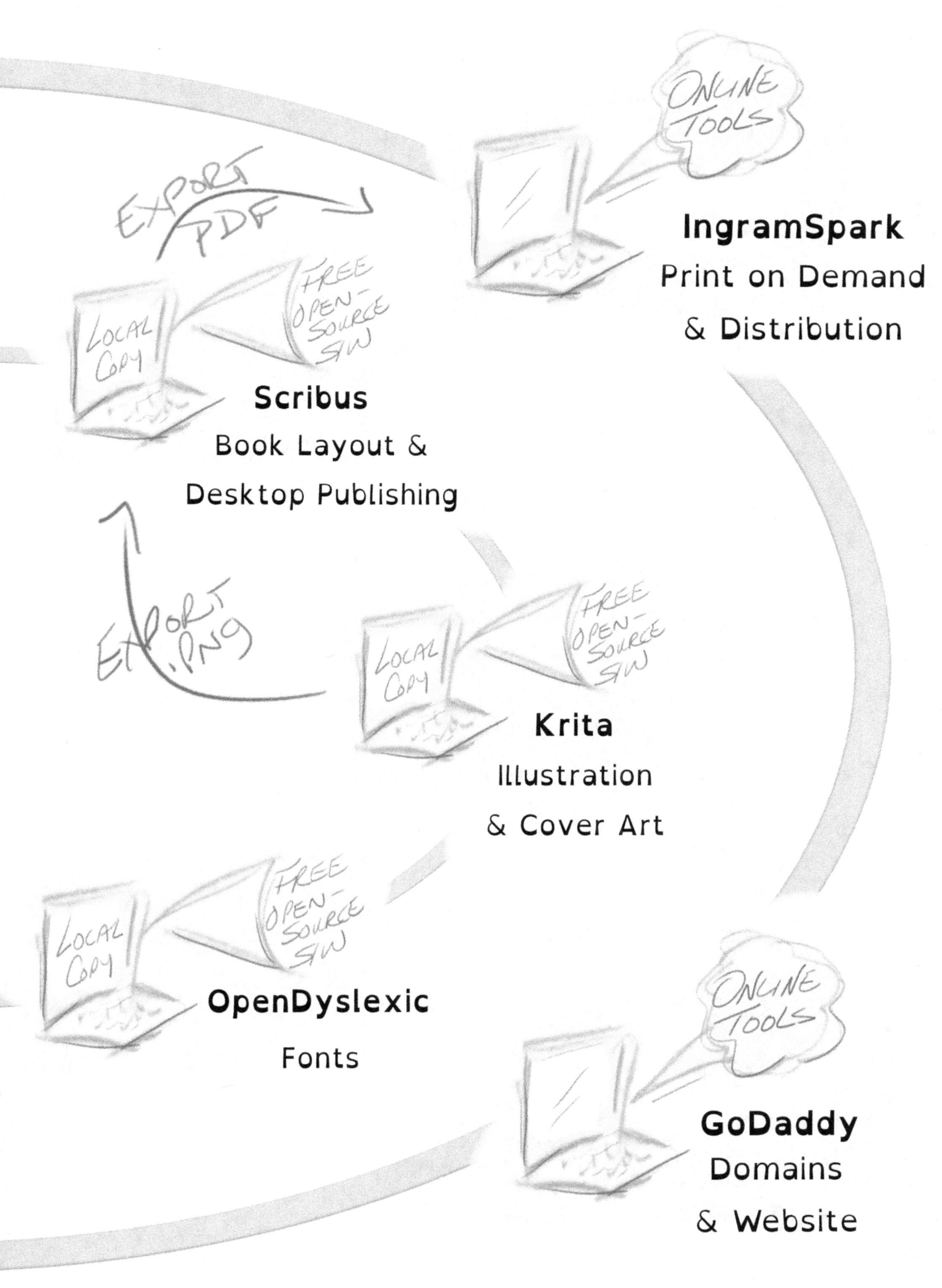

ONLINE TOOLS
EXPORT PDF
LOCAL COPY
FREE OPEN-SOURCE S/W
IngramSpark
Print on Demand
& Distribution
Scribus
Book Layout &
Desktop Publishing
EXPORT .PNG
LOCAL COPY
FREE OPEN-SOURCE S/W
Krita
Illustration
& Cover Art
LOCAL COPY
FREE OPEN-SOURCE S/W
OpenDyslexic
Fonts
ONLINE TOOLS
LOCAL COPY
GoDaddy
Domains
& Website